COLOR BY NUMBER

FOR GIRLS

UNICORNS, MERMAIDS, PRINCESSES & FAIRIES

AGES 4-8

MATH + COLORING = FUN LEARNING!

THIS BOOK BELONGS TO

_____

PRACTICE COLORING INSIDE THE LINES

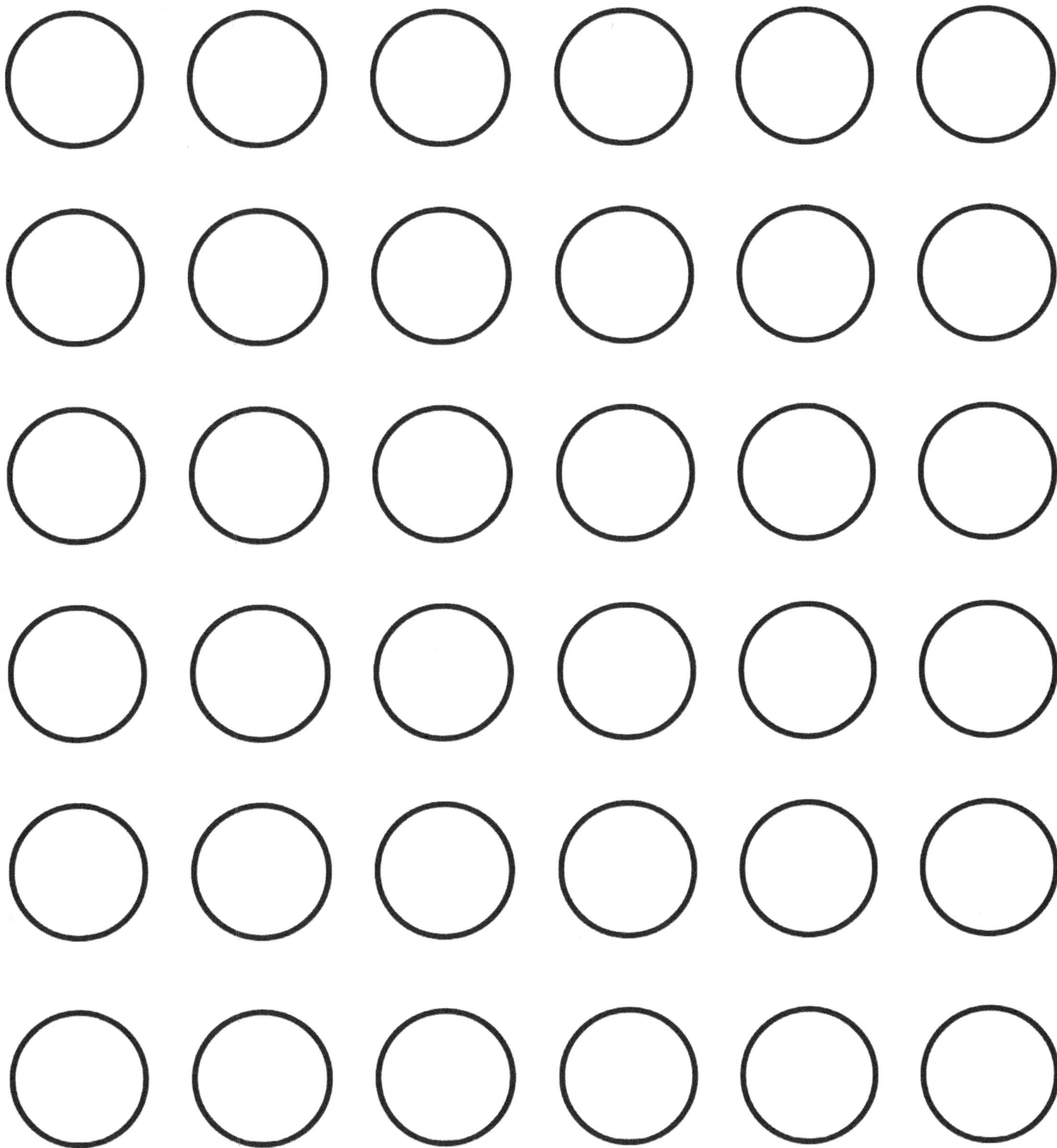

COLOR BY NUMBER: ADDITION

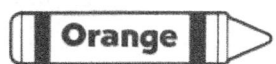

$0 + 1 =$

$1 + 1 =$

$1 + 2 =$

$1 + 3 =$

$1 + 4 =$

$1 + 5 =$

$1 + 6 =$

$1 + 7 =$

$1 + 8 =$

$1 + 9 =$

COLOR BY NUMBER: SUBTRACTION

 Light blue ▷ 00-03
 Pink ▷ 04-06
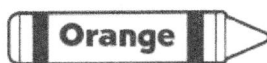 Orange ▷ 07-09
 Blue ▷ 10-12

 Violet ▷ 13-15
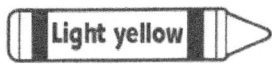 Light yellow ▷ 16-18
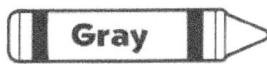 Gray ▷ 19-21
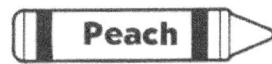 Peach ▷ 22-25

$6 - 6 = $

$12 - 11 = $

$16 - 2 = $

$7 - 1 = $

$2 - 1 = $

$8 - 4 = $

$5 - 1 = $

$8 - 2 = $

$5 - 1 = $

$16 - 4 = $

COLOR BY NUMBER: MULTIPLICATION

Light blue ▷ 00-03 Pink ▷ 04-06 Orange ▷ 07-09 Blue ▷ 10-12

Violet ▷ 13-15 Light yellow ▷ 16-18 light pink ▷ 19-21 Dark blue ▷ 22-25

2x10   10x2   21x1   7x3

21x1   3x3   19x1   5x3   10x2

3x3   4x2   7x1   3x3   3x6

4x2   9x2   4x4

8x1   3x6

9x2   3x5

3x3   2x10   3x1   2x1

9x1   1x1

3x3   7x1

2x10   4x5   8x1   4x1   3x3

9x1   1x2

7x3   7x3   2x1   1x2

7x2   6x1   1x2

3x2   9x1   5x1   9x1   1x1   1x2

4x1   3x5   7x3

1x2   3x1

21x1   1x3   2x10

6x3   9x2

19x1

$3 \times 1 =$

$5 \times 1 =$

$7 \times 1 =$

$9 \times 3 =$

$11 \times 1 =$

$13 \times 1 =$

$15 \times 5 =$

$17 \times 1 =$

$19 \times 1 =$

$21 \times 7 =$

COLOR BY NUMBER: DIVISION

| Light blue | 00-03 | Pink | 04-06 | Orange | 07-09 | Blue | 10-12 |
| Violet | 13-15 | Light yellow | 16-18 | Green | 19-21 | Peach | 22-25 |

$20\div2$ $20\div2$ $36\div3$

$22\div2$ $14\div2$ $3\div1$

$24\div2$ $24\div2$ $48\div4$

$2\div1$ $3\div1$ $1\div3$

$32\div2$ $1\div1$ $44\div2$

$8\div2$ $6\div2$ $20\div2$ $40\div4$

$12\div2$ $32\div2$

$32\div2$ $44\div2$ $6/2$

$36\div2$ $50\div2$ $24\div2$

$30\div3$ $36\div2$ $48\div4$ $26\div2$ $36\div2$ $22\div2$

$4\div2$ $12\div2$ $8\div2$

$40\div4$ $26\div2$ $50\div2$

$48\div4$ $30\div3$ $20\div2$

$44\div2$ $8\div2$ $12\div2$

$30\div3$ $20\div2$ $30\div2$ $10\div2$

$26\div2$ $48\div4$

$20\div2$ $12\div3$ $18\div3$ $12\div3$ $8\div2$

$24\div2$

$8\div2$ $30\div2$ $28\div2$ $12\div3$ $24\div4$ $40\div4$

$12\div2$ $26\div2$ $18\div3$

$40\div4$

$48\div4$ $16\div4$ $24\div4$ $28/2$ $16\div4$

$40\div2$ $30/2$ $12\div2$

$26\div2$

$42\div2$ $38\div2$ $57\div3$

$1 \div 1 = $ ___

$6 \div 3 = $ ___

$3 \div 1 = $ ___

$4 \div 2 = $ ___

$6 \div 1 = $ ___

$6 \div 3 = $ ___

$14 \div 2 = $ ___

$8 \div 4 = $ ___

$9 \div 3 = $ ___

$10 \div 5 = $ ___

COLOR BY NUMBER: ADDITION

 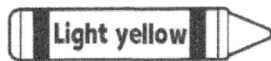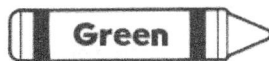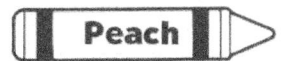

$1 + 1 =$

$2 + 1 =$

$3 + 2 =$

$4 + 3 =$

$5 + 4 =$

$6 + 5 =$

$7 + 6 =$

$8 + 7 =$

$9 + 8 =$

$10 + 9 =$

COLOR BY NUMBER: ADDITION

 Peach 00-03    Pink 04-06    Orange 07-09    Blue 10-12

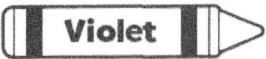 Violet 13-15   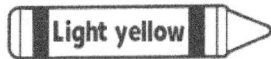 Light yellow 16-18   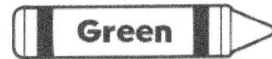 Green 19-21    Light blue 22-25

$2 + 2 =$ _____

$4 + 4 =$ _____

$6 + 5 =$ _____

$8 + 7 =$ _____

$10 + 9 =$ _____

$12 + 6 =$ _____

$14 + 10 =$ _____

$16 + 8 =$ _____

$12 + 5 =$ _____

$10 + 9 =$ _____

COLOR BY NUMBER: MULTIPLICATION

 Light blue 00-03    Pink 04-06   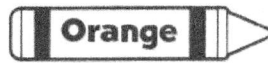 Orange 07-09    Blue 10-12

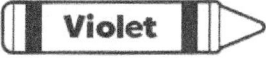 Violet 13-15   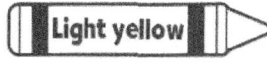 Light yellow 16-18    Green 19-21    Peach 22-25

$10 \times 2 = \underline{}$

$4 \times 2 = \underline{}$

$1 \times 1 = \underline{}$

$15 \times 3 = \underline{}$

$7 \times 1 = \underline{}$

$2 \times 2 = \underline{}$

$6 \times 3 = \underline{}$

$8 \times 4 = \underline{}$

$3 \times 2 = \underline{}$

$9 \times 9 = \underline{}$

COLOR BY NUMBER: DIVISION

Light blue ▷ 00-03 Pink ▷ 04-06 Orange ▷ 07-09 Blue ▷ 10-12
Violet ▷ 13-15 Light yellow ▷ 16-18 Green ▷ 19-21 Peach ▷ 22-25

20÷2   30÷3   48÷4   16÷4   22÷2   20/2   40÷4

24÷2   32÷2   34÷2   48÷4

22÷2   38÷2   36÷3   50÷5

50÷5   48÷3   4÷2   6÷2

2÷1   3÷1   30÷3

40÷4   44÷2

48÷4   33÷3   20÷2

30÷3   46÷2   50÷2

36÷3   50÷5   12÷4   36÷3   22÷2

36÷3   20÷2   6÷2   30÷3   48÷4

2÷1   44÷2   12÷6

24÷2   30÷2   26÷2   3÷1

20÷2   44÷2   28÷2   46÷2   20÷2   50÷5

48÷4   7÷1   8÷1   9÷1   14÷2   44÷4

16÷2   18÷2   21÷3   24÷3   36÷3

22÷2   40÷4   24÷2

27÷3   16÷2   18÷2   36÷3

14÷2

44÷4   1÷1   3÷3   22÷2   20÷2   40÷4
33÷3   36÷3   50÷5

$2 \div 1 =$

$4 \div 2 =$

$6 \div 3 =$

$8 \div 4 =$

$10 \div 5 =$

$12 \div 6 =$

$14 \div 7 =$

$16 \div 8 =$

$18 \div 9 =$

$20 \div 10 =$

COLOR BY NUMBER: SUBTRACTION

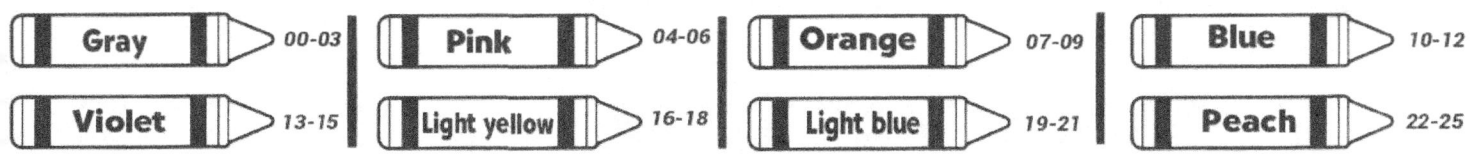

Gray 00-03

Pink 04-06

Orange 07-09

Blue 10-12

Violet 13-15

Light yellow 16-18

Light blue 19-21

Peach 22-25

19-4

18-4

18-3

18-4

20-4

10-1

18-2

20-13

17-2

20-7

18-2

15-7

20-6

18-3

15-6

25-3

8-1

19-4

3-1

17-2

10-7

20-6

18-3

20-6

27-3

10-4

18-3

29-7

15-2

17-2

1-1

10-5

1-1

18-4

18-4

11-5

19-4

20-7

19-4

28-5

2

19-6

20-6

18-3

29-7

19-4

20-6

19-5

19-4

25-4

19-5

20-7

20-6

21-2

20 - 4 =

24 - 5 =

21 - 2 =

21 - 3 =

16 - 4 =

10 - 5 =

10 - 2 =

24 - 8 =

16 - 4 =

22 - 11 =

COLOR BY NUMBER: SUBTRACTION

Light blue ▷ 00-03   Pink ▷ 04-06   Orange ▷ 07-09   Blue ▷ 10-12

Violet ▷ 13-15   Light yellow ▷ 16-18   Green ▷ 19-21   Peach ▷ 22-25

15-2   17-4   17-3   17-2

15-0   17-2

15-1   18-4

16-3   13-2   5-2   19-4

10-2   6-1

7-1   20-1

16-1   25-1   18-3   19-6

16-2   0-3

30-5   20-5   18-3   19-6

30-6   19-5

30-8   19-4

20-2   30-7   17-4

30-5   20-7   20-6

18-1

18-5   26-2   27-2   18-3

18-3   20-3   20-2   20-2   15-2   20-5

17-3

17-1   18-1

21-3   17-1

19-4   15-0   17-1   19-5

15-2

20-6   17-1   18-1   18-4

20-7   20-5   19-4

$$1 - 1 = $$

$$14 - 7 = $$

$$5 - 1 = $$

$$6 - 3 = $$

$$10 - 2 = $$

$$15 - 5 = $$

$$18 - 6 = $$

$$7 - 7 = $$

$$8 - 4 = $$

$$20 - 3 = $$

COLOR BY NUMBER: SUBTRACTION

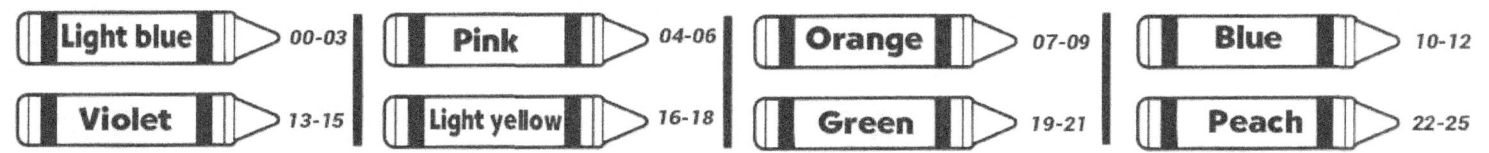

Light blue	00-03	Pink	04-06	Orange	07-09	Blue	10-12
Violet	13-15	Light yellow	16-18	Green	19-21	Peach	22-25

17-2
19-4
11-6
6-6
20-6
20-7

18-4
6-3
17-3

6-4
20-2

20-3
25-4
6-5

5-1
28-7
6-6
10-5
10-4

17-1
30-5

21-7
10-1
25-3

22-8
11-2
10-9

10-5
10-4

24-12
26-10
10-3
10-10

18-1
11-4
25-8

6-3
7-0

21-3
9-1
20-3

8-1

6-4
10-9

17-1

6-5
10-10

6-6

$4 - 1 =$ ____

$6 - 3 =$ ____

$10 - 2 =$ ____

$8 - 4 =$ ____

$12 - 3 =$ ____

$2 - 2 =$ ____

$16 - 4 =$ ____

$18 - 9 =$ ____

$6 - 6 =$ ____

$9 - 3 =$ ____

COLOR BY NUMBER: MULTIPLICATION

Light blue ▷ 00-03 Pink ▷ 04-06 Orange ▷ 07-09 Blue ▷ 10-12

Violet ▷ 13-15 Light yellow ▷ 16-18 Green ▷ 19-21 Peach ▷ 22-25

2x2   2x3   4x1   5x1

7x2   3x5   7x2   15x1

8x2   9x2   8x2   9x2

15x1   13x1   6x2   13x1   6x1   5x3

16x1   5x1   3x2   4x1   5x1   16x1

13x1   4x1   7x2

7x2   6x2   13x1   7x2

4x1   7x2   6x2   22x1   15x1   5x3   4x1

3x1   2x1   23x1   3x1   5x3

15x1   5x3   11x2   13x1   3x1   12x2   13x1

16x1   13x1   1x1   2x1   7x2   7x2   8x2

13x1   3x1   0x1

3x0   3x1   3x0

1x1   2x1   3x1   1x1   16x1

17x1   1x1

3x1   3x0

0x1   3x3

7x1

10x2   19x1   4x2   21x1   3x7

$0 \times 1 = $

$1 \times 1 = $

$1 \times 2 = $

$1 \times 3 = $

$1 \times 4 = $

$1 \times 5 = $

$1 \times 6 = $

$1 \times 7 = $

$1 \times 8 = $

$1 \times 9 = $

COLOR BY NUMBER: SUBTRACTION

Gray ▷ 00-03 Pink ▷ 04-06 Orange ▷ 07-09 Blue ▷ 10-12

Violet ▷ 13-15 Light blue ▷ 16-18 Green ▷ 19-21 Peach ▷ 22-25

25-9 25-8 25-7 24-8 18-6 24-7 24-6

19-1 4-1 16-7 14-7 3-1 23-7

4-2 15-8 10-6 14-6 23-6 3-2

14-5 13-4 12-5 23-5 22-5

19-2 13-5 22-6

13-6

19-3

25-0 20-7

20-6

10-3 17-5 19-5 22 21-4

20-2 20-5 25-1

25-2

20-3 12-4 21-5

10-2 18-5 12-3 21-3

20-4 30-5 10-1

21-3 20-4 10-6 10-5 20-4

19-8 19-7

30-8 20-4 10-4 20-3 30-7

15-3

21-4 23-5 20-10 19-9 20-3

20-8

20-9 19-3

18-7

19-8 20-2

21-5 18-8 19-7

18-6 19-2

16-6

22-5 17-7 17-6 17-5 19-1

22-6 22-7

$0 - 0 =$ ⬜

$1 - 0 =$ ⬜

$5 - 2 =$ ⬜

$9 - 3 =$ ⬜

$4 - 4 =$ ⬜

$6 - 5 =$ ⬜

$6 - 6 =$ ⬜

$10 - 7 =$ ⬜

$18 - 8 =$ ⬜

$21 - 9 =$ ⬜

COLOR BY NUMBER: MULTIPLICATION

| Light yellow | 00-03 | Pink | 04-06 | Orange | 07-09 | Blue | 10-12 |
| Violet | 13-15 | Light blue | 16-18 | Green | 19-21 | Peach | 22-25 |

$3 \times 7 =$

$2 \times 9 =$

$5 \times 2 =$

$6 \times 3 =$

$7 \times 1 =$

$2 \times 5 =$

$3 \times 3 =$

$5 \times 5 =$

$9 \times 2 =$

$1 \times 9 =$

COLOR BY NUMBER: ADDITION

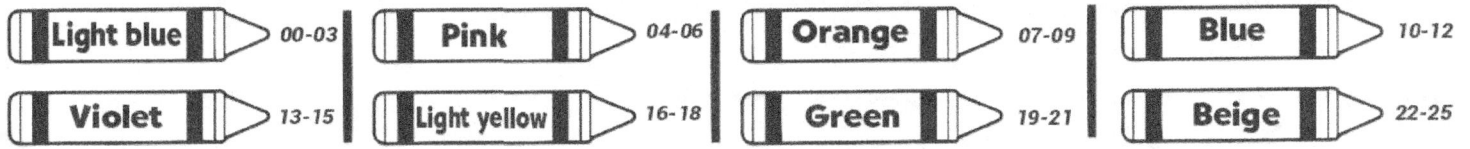

| Light blue | 00-03 | Pink | 04-06 | Orange | 07-09 | Blue | 10-12 |
| Violet | 13-15 | Light yellow | 16-18 | Green | 19-21 | Beige | 22-25 |

17+1 16+2 15+2 15+3 14+4 13+5 12+6

9+9

8+0 7+5 11+7 10+8

14+2 6+6 10+2

10+9

9+7 18+1 17+4 10+5 15+1

15+6 14+2

6+6 15+3 1+1 1+2 20+5 15+6 25+0 20+2

11+10

15+1 8+10 0+1 10+9 24+1 8+10

9+6

10+6 3+3 10+12 23+2

1+1 6+3 22+1 22+1 10+6

10+7 13+2 15+8 14+4

5+1 18+7 20+2 8+4

11+6 18+7 9+7

2+2 16-18 10+6 9+3

9+9 11+1 8+8

10+2 11+6

12+6 13+5 11+6

17+1 14+4 15+2

16+2 5+11 10+7

10 + 1 =

4 + 1 =

0 + 2 =

9 + 3 =

8 + 4 =

7 + 5 =

2 + 6 =

4 + 7 =

5 + 8 =

6 + 9 =

COLOR BY NUMBER: SUBTRACTION

 **Light blue** 00-03  **Pink** 04-06  **Orange** 07-09  **Blue** 10-12

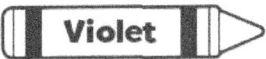 **Violet** 13-15 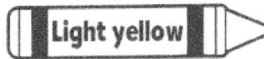 **Light yellow** 16-18 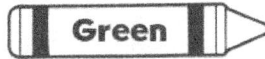 **Green** 19-21  **Peach** 22-25

$12 - 5 =$ _____

$20 - 9 =$ _____

$17 - 7 =$ _____

$13 - 10 =$ _____

$12 - 11 =$ _____

$8 - 5 =$ _____

$15 - 3 =$ _____

$7 - 3 =$ _____

$13 - 8 =$ _____

$15 - 1 =$ _____

COLOR BY NUMBER: MULTIPLICATION

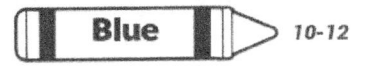

Color	Range
Light blue	00-03
Pink	04-06
Orange	07-09
Blue	10-12
Violet	13-15
Light yellow	16-18
Green	19-21
Peach	22-25

$0 \times 23 = $

$1 \times 25 = $

$2 \times 2 = $

$3 \times 3 = $

$4 \times 4 = $

$5 \times 5 = $

$4 \times 6 = $

$3 \times 7 = $

$2 \times 8 = $

$2 \times 9 = $

COLOR BY NUMBER: MULTIPLICATION

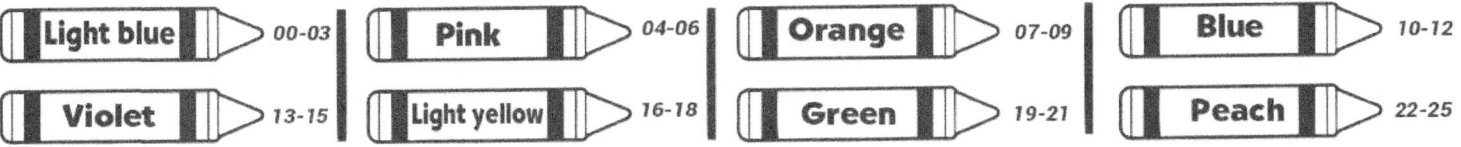

Light blue 00-03   Pink 04-06   Orange 07-09   Blue 10-12

Violet 13-15   Light yellow 16-18   Green 19-21   Peach 22-25

4 × 5 =

6 × 3 =

11 × 1 =

8 × 2 =

7 × 3 =

9 × 2 =

21 × 1 =

20 × 1 =

2 × 8 =

19 × 1 =

COLOR BY NUMBER: MULTIPLICATION

Pink 00-03   Light blue 04-06   Orange 07-09   Blue 10-12

Violet 13-15   Light yellow 16-18   Green 19-21   Peach 22-25

2x2   3x2   4x1   2x2

3x2   1x3   5x1   6x1

10x1   11x1   3x2

5x2   11x2   7x1   8x1

6x2   8x2

6x1   12x1   3x5   5x1

dot

5x1   9x1   0x0   4x1

22x1   3x1   0x1   2x1

8x1   3x3   9x1   8x3

5x5   3x8

4x1   11x2   3x8   1x1

3x2   7x1

5x3   2x7   22x1   3x5

2x2   23x1

3x2   6x3   4x1

2x2   2x7   2x8   9x2   6x3   2x2

3x5   1x13

4x1

5 × 5 =

6 × 4 =

11 × 2 =

12 × 1 =

7 × 3 =

10 × 2 =

9 × 2 =

2 × 7 =

15 × 1 =

16 × 1 =

COLOR BY NUMBER: SUBTRACTION

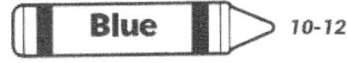

| Light blue | 00-03 | Pink | 04-06 | Orange | 07-09 | Blue | 10-12 |
| Violet | 13-15 | Light yellow | 16-18 | Green | 19-21 | Peach | 22-25 |

1 - 0 =

2 - 1 =

3 - 2 =

4 - 3 =

5 - 4 =

6 - 5 =

7 - 6 =

8 - 7 =

9 - 8 =

10 - 9 =

COLOR BY NUMBER: ADDITION

 Light blue 00-03
 Pink 04-06
 Orange 07-09
 Blue 10-12
 Violet 13-15
 Light yellow 16-18
 Green 19-21
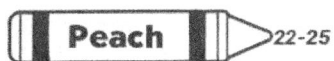 Peach 22-25

$7 + 1 = \square$

$5 + 1 = \square$

$3 + 12 = \square$

$1 + 3 = \square$

$9 + 14 = \square$

$11 + 5 = \square$

$5 + 16 = \square$

$3 + 7 = \square$

$7 + 10 = \square$

$9 + 9 = \square$

COLOR BY NUMBER: MULTIPLICATION

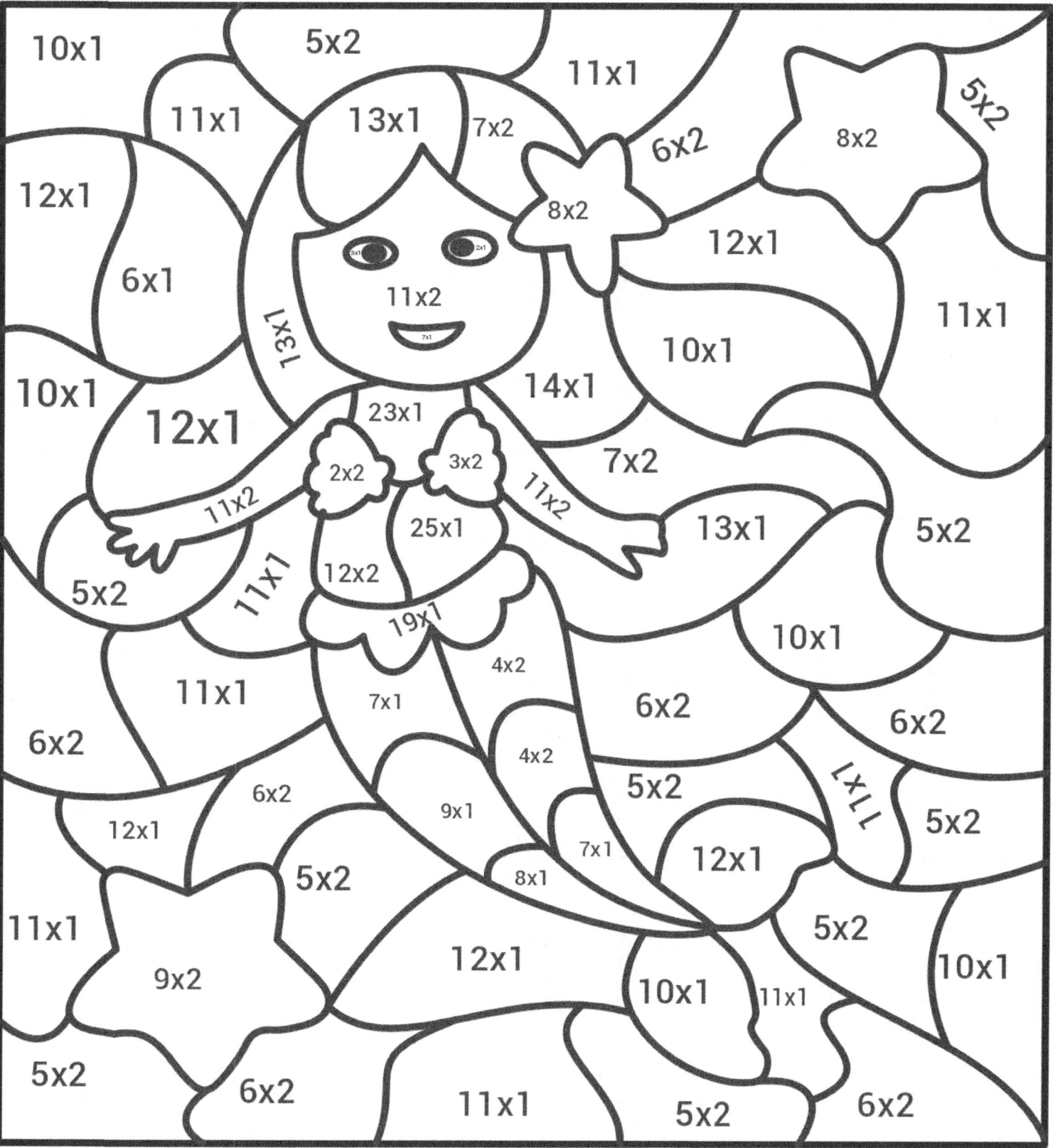

Blue ▷ 00-03   Pink ▷ 04-06   Orange ▷ 07-09   Light blue ▷ 10-12

Violet ▷ 13-15   Light yellow ▷ 16-18   Green ▷ 19-21   Peach ▷ 22-25

10x1   5x2   11x1   8x2   5x2

11x1   13x1   7x2   6x2

12x1   8x2

6x1   12x1

11x2   11x1

13x1   7x1   10x1

10x1   2x1   14x1

12x1   23x1   7x2

2x2   3x2   11x2

11x2   25x1   13x1   5x2

5x2   11x1   12x2   19x1

4x2   10x1

11x1   7x1   6x2   6x2

6x2   4x2   5x2   11x1

6x2   9x1   5x2

12x1   7x1   12x1

11x1   5x2   8x1   10x1

9x2   12x1   5x2

10x1   11x1   10x1

5x2   6x2   11x1   5x2   6x2

$22 \times 1 = $

$11 \times 1 = $

$5 \times 2 = $

$8 \times 3 = $

$6 \times 4 = $

$19 \times 1 = $

$18 \times 1 = $

$8 \times 3 = $

$3 \times 8 = $

$1 \times 9 = $

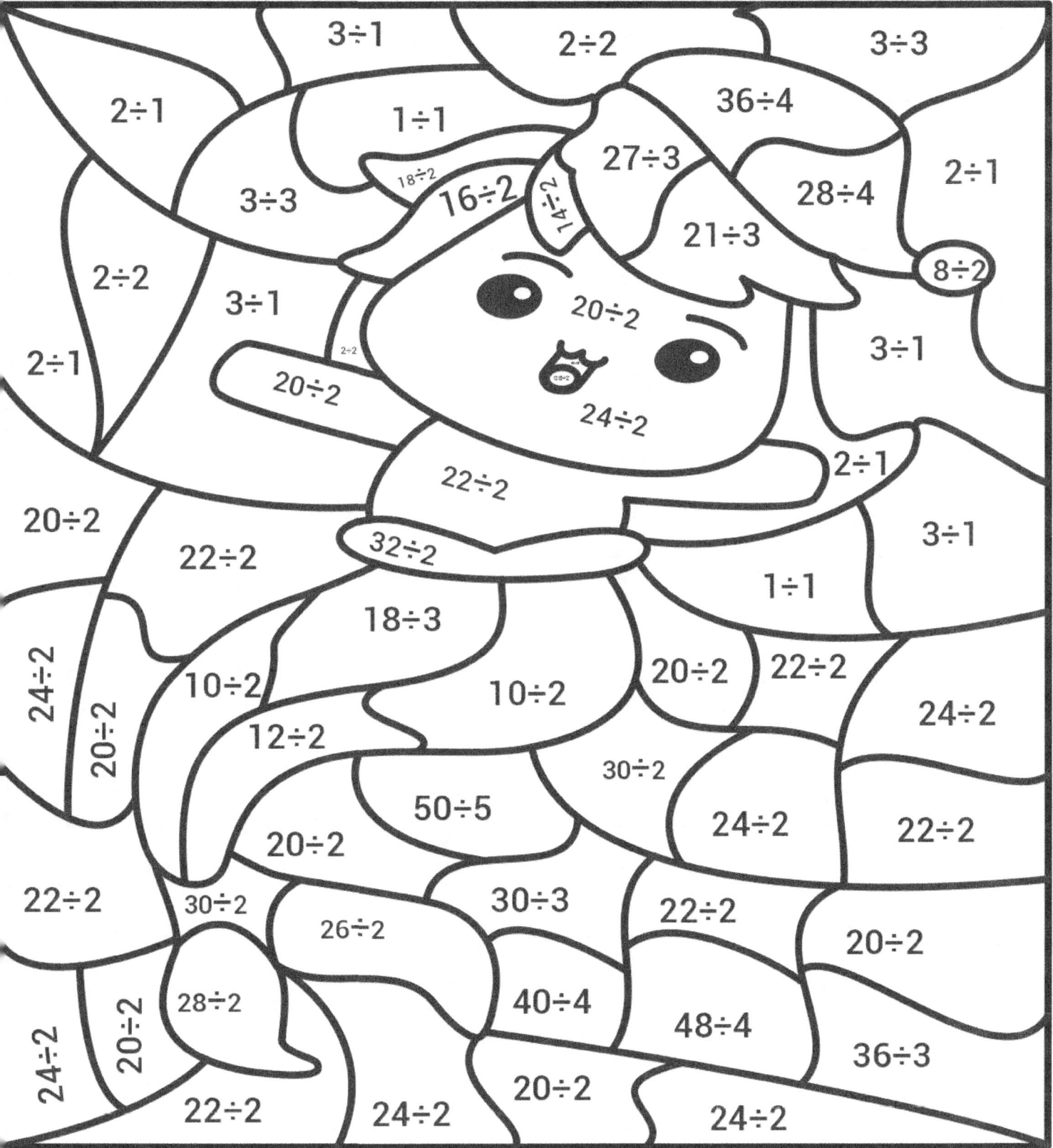

COLOR BY NUMBER: DIVISION

Light blue | 00-03
Pink | 04-06
Orange | 07-09
Blue | 10-12
Violet | 13-15
Light yellow | 16-18
Brown | 19-21
Peach | 22-25

$3 \div 1$

$2 \div 2$

$3 \div 3$

$2 \div 1$

$1 \div 1$

$36 \div 4$

$27 \div 3$

$2 \div 1$

$18 \div 2$

$16 \div 2$

$14 \div 2$

$28 \div 4$

$3 \div 3$

$21 \div 3$

$2 \div 2$

$3 \div 1$

$20 \div 2$

$8 \div 2$

$2 \div 2$

$3 \div 1$

$2 \div 1$

$20 \div 2$

$24 \div 2$

$22 \div 2$

$2 \div 1$

$20 \div 2$

$32 \div 2$

$3 \div 1$

$22 \div 2$

$1 \div 1$

$24 \div 2$

$18 \div 3$

$10 \div 2$

$10 \div 2$

$20 \div 2$

$22 \div 2$

$20 \div 2$

$12 \div 2$

$24 \div 2$

$30 \div 2$

$50 \div 5$

$24 \div 2$

$22 \div 2$

$20 \div 2$

$22 \div 2$

$30 \div 2$

$30 \div 3$

$22 \div 2$

$26 \div 2$

$20 \div 2$

$20 \div 2$

$28 \div 2$

$40 \div 4$

$48 \div 4$

$24 \div 2$

$20 \div 2$

$36 \div 3$

$22 \div 2$

$24 \div 2$

$20 \div 2$

$24 \div 2$

$0 \div 1 = $

$1 \div 1 = $

$6 \div 2 = $

$9 \div 3 = $

$16 \div 4 = $

$10 \div 5 = $

$18 \div 2 = $

$21 \div 7 = $

$24 \div 8 = $

$18 \div 9 = $

COLOR BY NUMBER: MULTIPLICATION

Light blue	00-03	Pink	04-06	Orange	07-09	Blue	10-12
Violet	13-15	Yellow	16-18	Green	19-21	Peach	22-25

2 × 1 =

5 × 1 =

1 × 2 =

8 × 3 =

9 × 2 =

5 × 5 =

3 × 6 =

6 × 4 =

7 × 3 =

10 × 2 =

COLOR BY NUMBER: SUBTRACTION

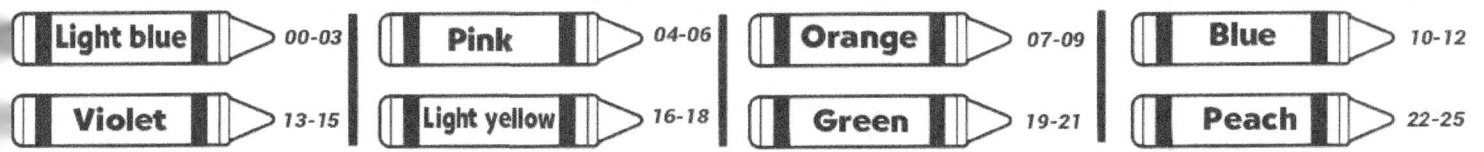

| Light blue | 00-03 | Pink | 04-06 | Orange | 07-09 | Blue | 10-12 |
| Violet | 13-15 | Light yellow | 16-18 | Green | 19-21 | Peach | 22-25 |

11 − 1 =

12 − 2 =

13 − 4 =

14 − 6 =

15 − 8 =

16 − 10 =

17 − 12 =

18 − 14 =

19 − 16 =

20 − 18 =

COLOR BY NUMBER: MULTIPLICATION

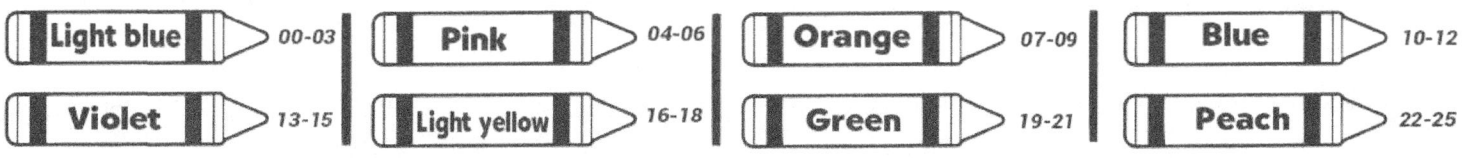

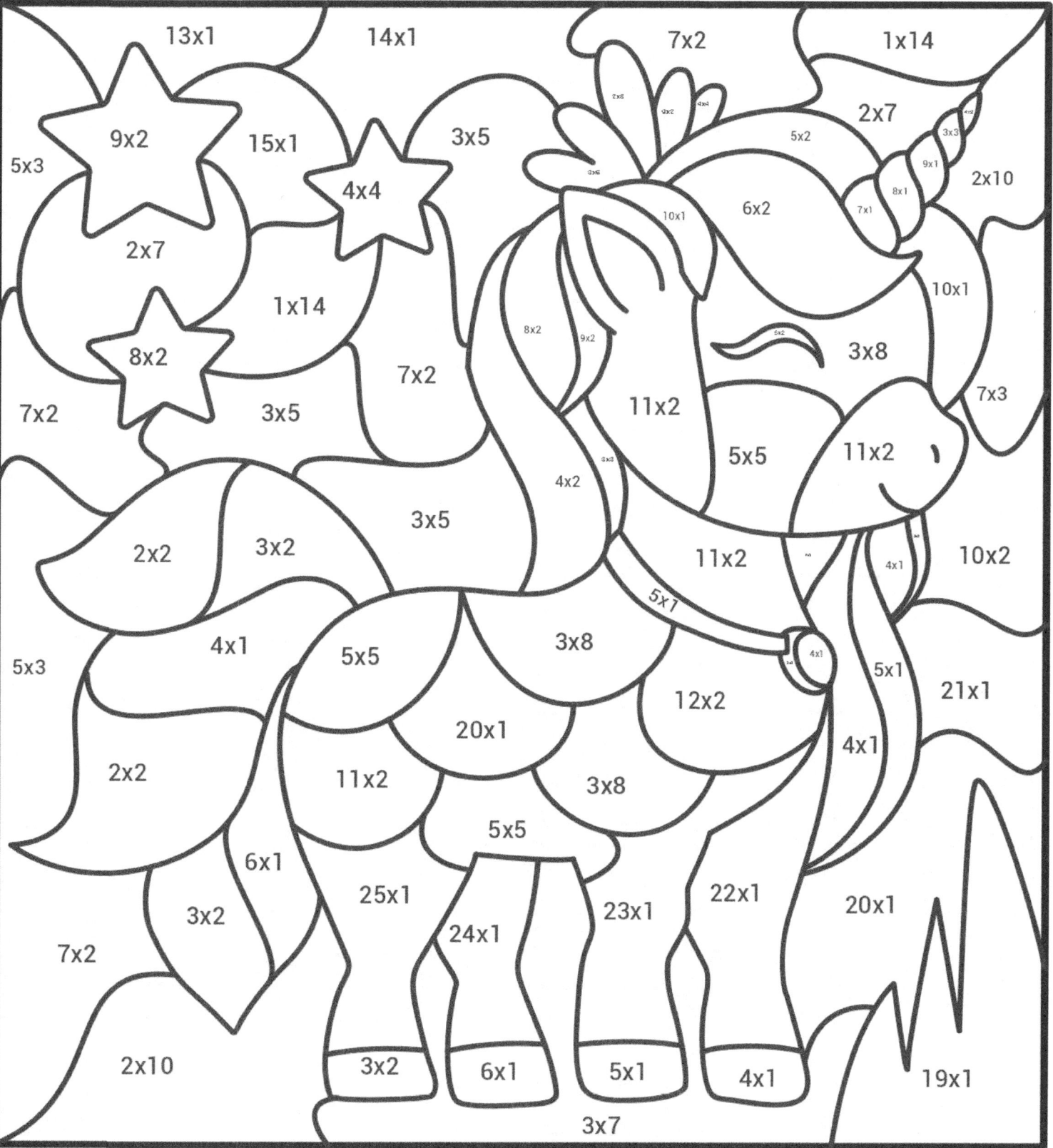

Color	Range
Light blue	00-03
Pink	04-06
Orange	07-09
Blue	10-12
Violet	13-15
Light yellow	16-18
Green	19-21
Peach	22-25

13x1　14x1　7x2　1x14

9x2　15x1　3x5　2x7　2x7

5x3　4x4　5x2　3x3

2x7　6x2　9x1

8x1　2x10

1x14　10x1　7x1

8x2　10x1

7x2　7x2　3x5　8x2　9x2　11x2　3x8

3x5　4x2　5x5　11x2　7x3

2x2　3x2　3x5　3x3　11x2　10x2

4x1　5x5　3x8　5x1　4x1

5x3　12x2　5x1　21x1

2x2　11x2　20x1　3x8　4x1

6x1　5x5　20x1

3x2　25x1　23x1　22x1

7x2　3x2　24x1

2x10　3x2　6x1　5x1　4x1　19x1

3x7

$6 \times 1 = \underline{}$

$6 \times 2 = \underline{}$

$1 \times 5 = \underline{}$

$4 \times 3 = \underline{}$

$7 \times 1 = \underline{}$

$5 \times 5 = \underline{}$

$8 \times 3 = \underline{}$

$3 \times 7 = \underline{}$

$2 \times 8 = \underline{}$

$1 \times 9 = \underline{}$

COLOR BY NUMBER: SUBTRACTION

8 − 4 =

9 − 7 =

10 − 9 =

11 − 3 =

12 − 6 =

13 − 5 =

14 − 7 =

15 − 10 =

16 − 15 =

17 − 3 =

COLOR BY NUMBER: ADDITION

25 + 10 =

24 + 9 =

23 + 11 =

22 + 12 =

21 + 4 =

20 + 15 =

19 + 19 =

18 + 17 =

17 + 8 =

16 + 9 =

COLOR BY NUMBER: ADDITION

Light blue ▷ 00-03
Pink ▷ 04-06
Orange ▷ 07-09
Blue ▷ 10-12
Violet ▷ 13-15
Light yellow ▷ 16-18
Green ▷ 19-21
Peach ▷ 22-25

10+7
5+11
17+1
12+6
13+5
5+1
10+5
4+0
4+2
4+0
2+2
4+2
4+2
5+3
6+2
3+3
3+1
4+2
8+1
11+6
20+2
1+5
14+4
6+1
16+2
1+5
11+4
22+1
6+1
4+2
23+2
4+2
3+3
6+1
1+5
0+1
1+5
7+2
15+2
2+0
4+1
0+0
4+2 20+2
2+0 3+3
5+4
7+2
15+2
0+0
2+0
15+2
7+1
4+4
16+2
0+0
1+2
2+0
11+6
13+5
2+0
3+0
1+1
3+0
12+6
1+1
0+1
17+1
1+2
15+2
14+4
12+6
14+4
17+1
13+5
24+1
11+6
20+5
1+2
16+2

2 + 1 =

2 + 1 =

2 + 2 =

2 + 3 =

2 + 4 =

2 + 5 =

2 + 6 =

2 + 7 =

2 + 8 =

2 + 9 =

COLOR BY NUMBER: ADDITION

Light yellow ➤ 00-03   Pink ➤ 04-06   Orange ➤ 07-09   Blue ➤ 10-12

Violet ➤ 13-15   Light blue ➤ 16-18   Green ➤ 19-21   Peach ➤ 22-25

10 + 11 =

9 + 10 =

8 + 12 =

7 + 13 =

6 + 15 =

5 + 17 =

4 + 19 =

3 + 17 =

2 + 18 =

0 + 19 =

COLOR BY NUMBER: ADDITION

 Brown ▷ 00-03
 Pink ▷ 04-06
 Orange ▷ 07-09
 Blue ▷ 10-12

 Violet ▷ 13-15
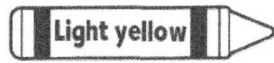 Light yellow ▷ 16-18
 Green ▷ 19-21
 Peach ▷ 22-25

5 + 12 =

5 + 13 =

5 + 14 =

5 + 15 =

5 + 16 =

5 + 17 =

5 + 18 =

5 + 19 =

5 + 20 =

5 + 15 =

COLOR BY NUMBER: ADDITION

- **Light blue** 00-03
- **Pink** 04-06
- **Orange** 07-09
- **Blue** 10-12
- **Violet** 13-15
- **Light yellow** 16-18
- **Green** 19-21
- **Peach** 22-25

1+2 3+0 2+0 1+0 1+1

7+6 2+2 2+1

1+1 7+8 7+7 9+7 3+0

1+0 8+8 8+10 1+1

2+0 8+5 8+9 11+14

9+9 9+8 8+10 1+1

3+0 9+7 8+10 8+9

1+1 11+13 1+1 2+1

8+10 11+12 7+8 8+8

1+2 9+7 9+8 2+0

8+9 10+7 9+9 11+11

8+8 10+6 6+2 1+2

2+0 6+6 10+8 6+1 5+4 3+0

6+4 6+5 2+0 1+1

5+7 5+3 5+2 2+1

5+6 1+2 1+0 1+0

5+5 4+4 4+5

1+0 1+1 2+1 4+3 1+0 1+2

3+0 2+0 1+1 2+0

7+3

24 + 1 =

23 + 1 =

22 + 2 =

21 + 3 =

20 + 4 =

19 + 5 =

18 + 6 =

15 + 7 =

13 + 8 =

11 + 9 =

COLOR BY NUMBER: SUBTRACTION

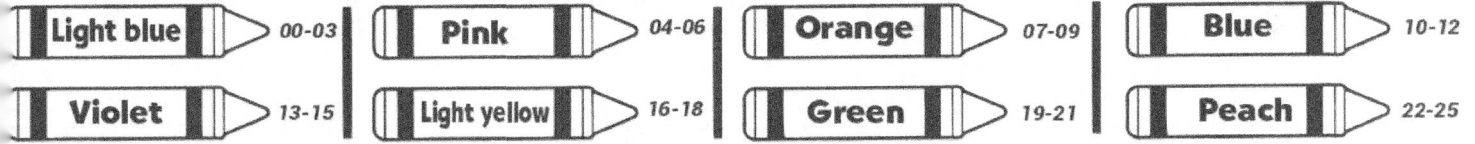

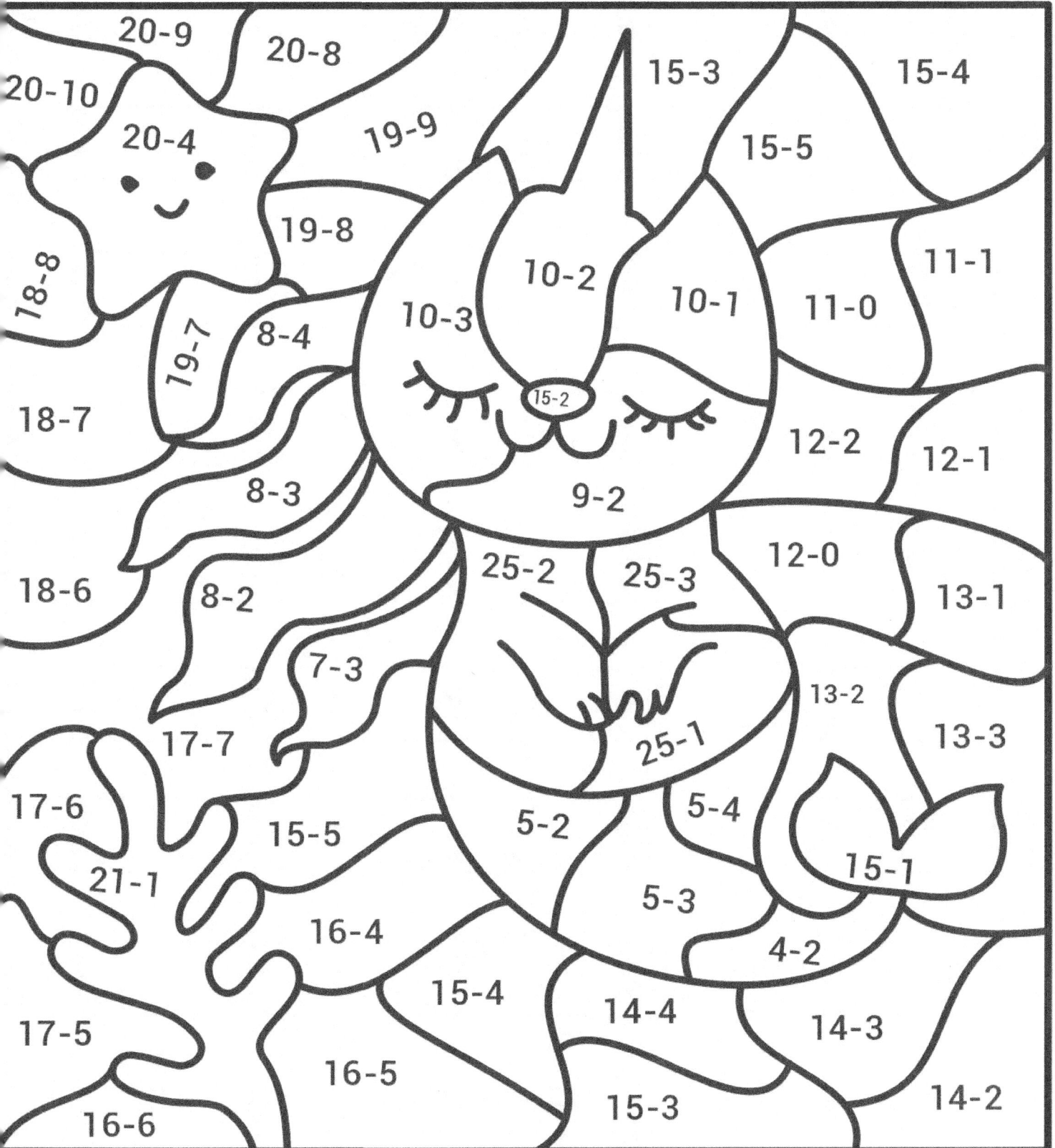

20-9   20-8   15-3   15-4

20-10   20-4   15-5

19-9

18-8   19-8   11-1

19-7   8-4   10-3   10-2   10-1   11-0

15-2

18-7   12-2   12-1

8-3   9-2

18-6   8-2   12-0   13-1

7-3   25-2   25-3

17-7   13-2   13-3

25-1

17-6   5-4

21-1   15-5   5-2   15-1

5-3

16-4   4-2

17-5   15-4   14-4   14-3

16-5

16-6   15-3   14-2

15 - 8 =

16 - 11 =

17 - 6 =

18 - 9 =

19 - 3 =

20 - 18 =

21 - 1 =

22 - 19 =

23 - 2 =

24 - 12 =

COLOR BY NUMBER: SUBTRACTION

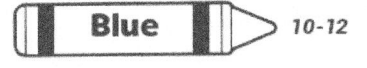

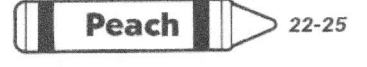

| Orange | 00-03 | Pink | 04-06 | Light blue | 07-09 | Blue | 10-12 |
| Violet | 13-15 | Light yellow | 16-18 | Green | 19-21 | Peach | 22-25 |

15 − 9 =

14 − 8 =

13 − 7 =

12 − 5 =

11 − 4 =

10 − 3 =

9 − 2 =

8 − 1 =

7 − 0 =

6 − 6 =

COLOR BY NUMBER: DIVISION

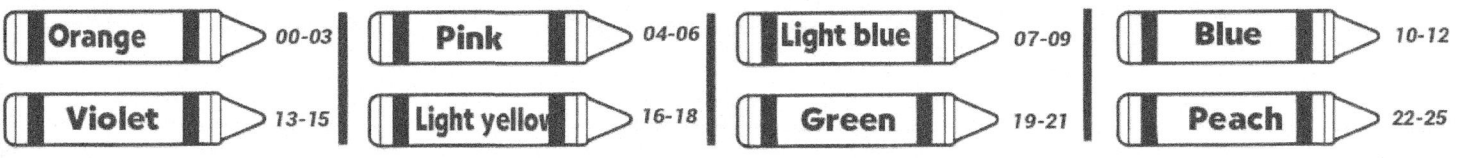

| Orange | 00-03 | Pink | 04-06 | Light blue | 07-09 | Blue | 10-12 |
| Violet | 13-15 | Light yellow | 16-18 | Green | 19-21 | Peach | 22-25 |

14÷2   18÷2   16÷2   18÷2

16÷2   26÷2   28÷2   24÷2

22÷3   44÷2   14÷2

24÷2   20÷2   24÷2   26÷2

22÷2

14÷2   26÷2

16÷2   28÷2   30÷2   28÷2   18÷2

22÷2   26÷2   44÷2   24÷2

30÷2   28÷2   28÷2   13-15   14÷2

24÷2   26÷2   26÷2

32÷2

16÷2   18÷2   34÷2   18÷2

40÷2   36÷2   16÷2

21÷2   18÷2   38÷2

14÷2   24÷2

24÷2   42÷2

18÷2   14÷2   16÷2   18÷2

$9 \div 3 =$ ____

$5 \div 5 =$ ____

$3 \div 1 =$ ____

$8 \div 2 =$ ____

$6 \div 3 =$ ____

$11 \div 1 =$ ____

$4 \div 4 =$ ____

$1 \div 1 =$ ____

$5 \div 1 =$ ____

$2 \div 1 =$ ____

COLOR BY NUMBER: MULTIPLICATION

Light blue ▷ 00-03 Pink ▷ 04-06 Orange ▷ 07-09 Blue ▷ 10-12

Violet ▷ 13-15 Light yellow ▷ 16-18 Green ▷ 19-21 Peach ▷ 22-25

$9 \times 1 =$ ____

$9 \times 1 =$ ____

$8 \times 2 =$ ____

$8 \times 2 =$ ____

$8 \times 3 =$ ____

$7 \times 1 =$ ____

$7 \times 2 =$ ____

$7 \times 3 =$ ____

$5 \times 4 =$ ____

$5 \times 5 =$ ____

COLOR BY NUMBER: ADDITION

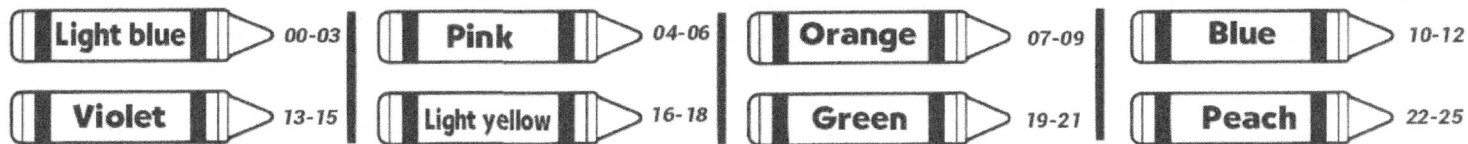

Color	Range
Light blue	00-03
Pink	04-06
Orange	07-09
Blue	10-12
Violet	13-15
Light yellow	16-18
Green	19-21
Peach	22-25

24 + 1 =

22 + 3 =

20 + 5 =

18 + 7 =

16 + 4 =

14 + 5 =

12 + 6 =

10 + 7 =

8 + 8 =

6 + 9 =

COLOR BY NUMBER: ADDITION

Light yellow 00-03	**Pink** 04-06	**Orange** 07-09	**Blue** 10-12
Violet 13-15	**Light blue** 16-18	**Gray** 19-21	**Peach** 22-25

10+7 11+6 12+6 16+2 15+2

14+4 15+2 0+0 5+1 5+11

13+5 10+2 1+1 4+2 17+1

5+11 0+1 10+7 0+0

11+1 2+0

17+1 20+2 3+0 10+7

11+6 10+12 18+7

7+3 1+2 10+12 11+6

11+6 15+6 18+7 20+2

11+10 22+1 13+5

18+1 0+1

16+2 1+2 12+6

14+4 1+1 14+4

24+1 12+6 23+2 17+1 16+2

13+5 15+2

11+6 10+7 5+11

25 + 13 =

21 + 1 =

18 + 2 =

12 + 3 =

13 + 4 =

14 + 5 =

10 + 6 =

11 + 7 =

16 + 8 =

17 + 9 =

COLOR BY NUMBER: MULTIPLICATION

Light blue ▷ 00-03 Pink ▷ 04-06 Orange ▷ 07-09 Blue ▷ 10-12

Violet ▷ 13-15 Light yellow ▷ 16-18 Green ▷ 19-21 Peach ▷ 22-25

5x2 6x2 5x2 11x1

10x1 6x2

12x1 2x1 1x1

10x1 11x1 3x4

3x1 11x2 12x2 11x1

3x4 1x1 11x2 6x2

2x1 5x2

3x1 2x1

2x1 2x1 3x1 10x1

5x2 2x1 3x1

12x2

6x2 1x1 11x1

10x1 5x2

9x2 3x2 5x2

7x2 7x2 6x2

2x2 3x2

10x2 8x2 5x1

8x2 9x2 2x2

5x1

11x1 5x2

11x1 5x1 3x2

3x4 6x2 5x2 2x2 3x2 5x1

11x1

$0 \times 1 =$

$25 \times 1 =$

$7 \times 2 =$

$5 \times 3 =$

$4 \times 4 =$

$4 \times 5 =$

$3 \times 6 =$

$3 \times 7 =$

$3 \times 8 =$

$1 \times 9 =$

COLOR BY NUMBER: SUBTRACTION

Light blue 00-03
Pink 04-06
Orange 07-09
light gray 10-12
Violet 13-15
Light yellow 16-18
Green 19-21
Peach 22-25

20-10
20-9
20-8
19-9
19-8
14-4
15-8
15-7
15-6
15-3
18-8
19-7
15-4
25-1
2-1
15-5
17-3
20-7
15-8 15-7
15-3
20-6
18-7
3-1
20-5
15-6 14-7
11-1
5-2 25-2 4-1
4-2
5-3 4-2
17-6
11-0
5-4 4-3
14-6 14-5
25-9
25-7
24-8
20-5
29-9
25-8
15-3
19-6
4-3
29-10
15-4
19-5
24-7
19-4
18-5
25-5
12-0
2-1
30-9
22-5
18-7
13-6
24-6 23-7
22-6
30-8
24-4
23-5
24-5
12-1
30-10
20-3
23-6
21-5 22-4
12-4
25-4
12-2
20-4
12-5
25-5
21-3
21-4
19-1
25-6
30-11
20-2
18-6
20-10
13-2
2-0
19-3
30-8
19-2
17-7
13-3
14-4
30-7
16-6
17-5
14-2
15-4
24-7
16-4
17-6
25-8
14-3
15-3
25-7 24-6
15-5
16-5

10 - 1 =

9 - 1 =

8 - 2 =

7 - 3 =

6 - 4 =

5 - 5 =

4 - 4 =

3 - 1 =

2 - 2 =

1 - 0 =

COLOR BY NUMBER: MULTIPLICATION

Light yellow 00-03 | Pink 04-06 | Orange 07-09 | Blue 10-12
Violet 13-15 | Light blue 16-18 | Green 19-21 | Peach 22-25

$0 \times 0 =$

$3 \times 1 =$

$5 \times 2 =$

$6 \times 3 =$

$2 \times 4 =$

$4 \times 5 =$

$9 \times 2 =$

$1 \times 7 =$

$7 \times 3 =$

$8 \times 2 =$

COLOR BY NUMBER: ADDITION

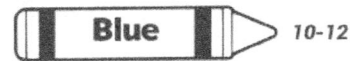

Light blue 00-03 Pink 04-06 Orange 07-09 Blue 10-12
Violet 13-15 Light yellow 16-18 Green 19-21 Peach 22-25

3 + 1 =

3 + 1 =

3 + 22 =

3 + 3 =

3 + 14 =

3 + 15 =

3 + 6 =

3 + 17 =

3 + 8 =

3 + 19 =

COLOR BY NUMBER: ADDITION

7 + 1 =

8 + 1 =

9 + 2 =

10 + 3 =

11 + 4 =

12 + 5 =

13 + 6 =

14 + 7 =

15 + 8 =

16 + 9 =

COLOR BY NUMBER: SUBTRACTION

| Red | 00-03 | Light blue | 04-06 | Orange | 07-09 | Green | 10-12 |
| Violet | 13-15 | Light yellow | 16-18 | Blue | 19-21 | Peach | 22-25 |

10 - 1 =

13 - 1 =

22 - 2 =

22 - 3 =

12 - 4 =

15 - 1 =

11 - 6 =

17 - 7 =

5 - 2 =

18 - 6 =

COLOR BY NUMBER: ADDITION

8+8 16+1 15+3 15+2 15+1 14+4

1+2 2+0 4+3 4+4 4+5 7+2 1+0

8+9 1+1 2+1 3+6 5+2 5+3 1+1 2+1 2+0 14+3

3+0 3+5 10+9 10+10 5+3 3+0 7+1

2+5 5+2 5+3

8+10 2+6 3+4 5+4 5+4 6+3

17+4 6+1 10+11 6+1 14+2

2+7 7+6 11+11 18+2 6+2

16+4 17+3 18+3 7+7 11+8 13+5

9+7 17+2 11+9

16+5 15+6

9+8 16+3 15+5 2+2 2+4 11+10 13+4

15+6 2+3 11+12 8+4 20+1 19+1 19+2 12+7 10+9

14+7 3+1 3+3

15+4 3+2 12+8 13+3

14+6 5+1 12+9

9+9 4+0 7+8 4+2 13+6

14+5 4+1 8+6 12+6

13+8 8+7 13+7

10+6 8+5 9+5 12+5

10+7 9+4 3+0

10+8 2+1 11+7 12+4

11+5 11+6

$0 + 1 =$ ___

$7 + 13 =$ ___

$7 + 11 =$ ___

$7 + 9 =$ ___

$7 + 7 =$ ___

$7 + 14 =$ ___

$7 + 12 =$ ___

$7 + 10 =$ ___

$7 + 17 =$ ___

$7 + 18 =$ ___

COLOR BY NUMBER: ADDITION

Light blue 00-03   Pink 04-06   Orange 07-09   Blue 10-12

Violet 13-15   Light yellow 16-18   Green 19-21   Peach 22-25

8+8   8+9   8+10   9+7

7+5   7+7   9+9

9+8   7+8   10+6

9+8   9+7   9+9   8+7   10+7

8+6   12+11   8+8   10+8

8+10   8+5   11+14   8+5   5+2   7+6

5+3   5+4   7+8   11+5

8+9   5+3   9+8   7+7

2+4   5+3   11+11

8+8   12+12   8+8   5+2   2+4   11+6

4+1   11+12   7+7

3+1   2+4   8+9   1+1   1+2   11+7

9+10

2+3   9+11   12+4   3+2

2+2   9+12   1+2   3+3

10+9   3+0   1+1

2+1

5+1   10+10   2+2   4+0

2+0   2+0

5+0   10+11   1+2   3+0   2+1   3+2

11+8   1+1   1+0   2+1

4+2   11+9   1+1   2+1

4+1   2+3   3+1

11+10   14+5

4+2   13+8   4+2   4+1

5+5   12+7   2+4   13+7

12+8   13+6   13+7   1+7

3+1   12+9   5+0   1+2

5+6   3+2   5+1

4+0   3+3   2+4   2+3   2+2   6+0

$$10 + 12 =$$

$$9 + 11 =$$

$$8 + 13 =$$

$$7 + 14 =$$

$$6 + 5 =$$

$$5 + 4 =$$

$$4 + 3 =$$

$$3 + 2 =$$

$$2 + 1 =$$

$$1 + 0 =$$

COLOR BY NUMBER: SUBTRACTION

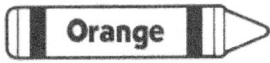 **Orange** *00-03*
 **Pink** *04-06*
 **Light blue** *07-09*
 **Blue** *10-12*

 **Violet** *13-15*
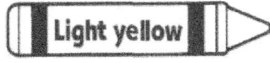 **Light yellow** *16-18*
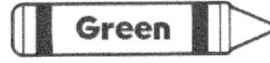 **Green** *19-21*
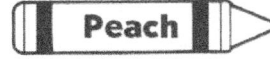 **Peach** *22-25*

11 − 11 =

12 − 12 =

13 − 12 =

14 − 11 =

4 − 2 =

5 − 5 =

6 − 6 =

9 − 7 =

11 − 8 =

15 − 9 =

COLOR BY NUMBER : DIVISION

Crayon	Range
Violet	00-03
Pink	04-06
Orange	07-09
Blue	10-12
Light blue	13-15
Light yellow	16-18
Light brown	19-21
Peach	22-25

$0 \div 0 = $ ___

$1 \div 1 = $ ___

$2 \div 2 = $ ___

$3 \div 3 = $ ___

$4 \div 4 = $ ___

$5 \div 5 = $ ___

$6 \div 6 = $ ___

$7 \div 7 = $ ___

$8 \div 8 = $ ___

$9 \div 9 = $ ___

COLOR BY NUMBER: ADDITION

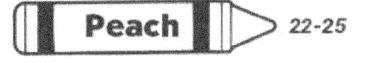

13+5   11+6   14+4   17+1   15+2

15+2   1+1   16+2   12+6   13+5

15+2   11+1

1+2   15+2   10+2   11+4   14+1   13+2   12+0

13+5   11+6

0+1   15+2   10+5   12+3   9+7   15+2

3+0   2+0   13+5   7+6   10+12   10+3   14+4

5+11   4+2   9+4   18+5   9+4   3+3   16+2

14+4   5+1   20+2

15+2   5+4   5+1

5+1   6+1   8+1   6+2   10+7

13+5   7+2   22+1   6+5   5+11

17+1   7+1   5+3   13+5

4+4   18+1

13+5   23+2   4+2   12+6

18+1   24+1   17+1

17+1   11+6   10+9

17+4   3+1   17+1   11+10   17+4

10+9   2+2

15+6   18+1   10+9   15+6   18+1

$3 + 1 = \underline{}$

$5 + 1 = \underline{}$

$7 + 2 = \underline{}$

$9 + 3 = \underline{}$

$11 + 4 = \underline{}$

$13 + 5 = \underline{}$

$15 + 6 = \underline{}$

$17 + 7 = \underline{}$

$18 + 8 = \underline{}$

$19 + 9 = \underline{}$

COLOR BY NUMBER: ADDITION

Light blue ▷ 00-03   Pink ▷ 04-06   Orange ▷ 07-09   Blue ▷ 10-12
Violet ▷ 13-15   Light yellow ▷ 16-18   Green ▷ 19-21   Beige ▷ 22-25

8 + 1 =

18 + 1 =

17 + 8 =

24 + 8 =

20 + 8 =

25 + 8 =

11 + 8 =

9 + 8 =

21 + 8 =

15 + 8 =

Made in the USA
Las Vegas, NV
30 June 2025